Bibliographic information published by the German National Library:

The German National Library lists this publication in the National Bibliography; detailed bibliographic data are available on the Internet at http://dnb.dnb.de .

Imprint:

Copyright © 2016 GRIN Verlag, Open Publishing GmbH
Print and binding: Books on Demand GmbH, Norderstedt Germany
ISBN: 9783668337077

This book at GRIN:

http://www.grin.com/en/e-book/343281/land-reform-in-zimbabwe-context-and-sypnosis

Harel Tanjong

Land Reform In Zimbabwe. Context and Sypnosis

GRIN Publishing

GRIN - Your knowledge has value

Since its foundation in 1998, GRIN has specialized in publishing academic texts by students, college teachers and other academics as e-book and printed book. The website www.grin.com is an ideal platform for presenting term papers, final papers, scientific essays, dissertations and specialist books.

Visit us on the internet:

http://www.grin.com/

http://www.facebook.com/grincom

http://www.twitter.com/grin_com

Land Reform in Zimbabwe

3/23/2016

Content

Introduction

A landlocked nation in Southern Africa, the area now known as the Republic of Zimbabwe has gone through many changes in the last century or so. From being under British rule at the turn of the twentieth century, to white minority rule, and finally a black African government. The story of this country is one of turbulent, and often violent, change. Today, the country faces many challenges, but one issue in particular stands out, an issue that has its roots in the very beginnings of the land as a British colony. The large land ownership by whites through most of the 20th century, and the subsequent land reform policies of the black government.

The land reform policies of the ZANU-PF government have contributed to a substantial downturn in the living standards of the Zimbabwean people. This paper provides historical context of the region's unequal land distribution. Also examined are the ZANU-PF government's' policies and attitudes from its assumption of power in the early 80's through the end of its "Fast Track" reform. The numerous economic effects of land reform, such as the country's collapse in agriculture, drop in living standards, and its notorious hyperinflation, will also be covered.

British Acquisition and Zimbabwean Rebellion

British Land Policies

Cecil Rhodes and his British South Africa Company (BSAC) first arrived on the lands of what is now Zimbabwe in the 1880's. Lavish mineral wealth existed in the area, particularly the gold fields of Mashonaland (Galbraith 1974). The British government, through the Company, encouraged mining firms, and regular citizens to come settle (Galbraith 1974). Initially, the Company intended to finance the governance of the colony through its mining exports alone, however, when gold and other mineral discoveries did not meet expectations, the administration was forced to find another source of revenue. . With its own expensive military and police force, the Company would have gone bankrupt if it was not for the huge agricultural potential of Rhodesia (Galbraith 1974).

Large-scale agriculture was quickly seen as the way to raise funds. In order to utilize this resource, the BSAC had to take over the land. A land commission was set up in 1894, and apportioned the prime land in the interior to British citizens. The Africans, who held 10 million acres before European arrival, were forced to the more arid periphery, crammed into about 2.5 million acres (Floyd 1962) "By the time of independence, in 1980, population densities were over three times greater in the black than in the white areas" (Palmer 1990, 165) With little land to cultivate, as well as being taxed by the Company, Africans had little choice but to work for the European-owned farms in the interior to make a living, of which their labor was in great demand (Rowe 2001).

In an effort to promote "conservation" in the urban areas, the government of the newly established Southern Rhodesia, a self-governing state, passed the Land Husbandry Act. The law, whose main goal was to "provide for a reasonable standard of good husbandry and for the protection of resources by all Africans using the land" (Zvobgo 2009, 56), limited stock holding by the natives to the carrying capacity of the land. Land ownership was also limited to eight acres, two for grazing and six for crops (Zvobgo 2009). Predictably, the Africans were vigorously opposed to the measure. To the average villager, cattle ownership was almost non-negotiable, and asking them to reduce their stock was a direct blow to their livelihood and wealth (Zvobgo 2009).

Black anger over land distribution reached a fever pitch in the 1960s, as the land currently known as Zimbabwe was then known as Rhodesia, and headed by white Briton, Ian Smith. Two black nationalist, left-wing organizations were established in the form of the Zimbabwe African National Union (ZANU) and the Zimbabwe African People's Union (ZAPU). Led by figures such as Ndabaningi Sithole and Robert Mugabe, as well as being funded by the Soviet Union and China, they fought a violent guerilla war against the government of Rhodesia in the 1970s (Palmer 1990).

Rhodesian Bush War

Irritated by Britain's stance of its colonies having no independence before majority rule, as well as a desire to strip the native Africans of the few remaining rights they had, the Rhodesian government declared independence from Britain in 1965 (Moorcraft 1990). From the perspective of the whites, their very livelihood was under attack. The white Rhodesians did not want to see the kind of instability and chaos that had occurred previously in Kenya and Congo (Moorcraft 1990). A perceived invasion of foreign funded communist triggered a hard shift to the political right amongst the white populace (Baxter 2016).

Rhodesia's Unilateral Declaration of Independence from Britain signaled to the black nationalist groups that they needed to act quickly to achieve liberation. Their strategy would rely on setting up camps in neighboring Zambia and Tanzania and invading Guerilla style into Rhodesian territory (Baxter 2016). Initially the superior efficiency of the Rhodesian government's armed forces, as well as its intelligence service, blunted all insurgent attacks (Baxter 2016). However, the ZANU and ZAPU rebels turned to the outside for assistance and training, receiving it from China and the Soviet Union (Baxter 2016). The foreign support for the nationalist, the new front that opened on the border with Mozambique, and the duration of the war, combined to eventually cripple the Rhodesian government's war effort. The two insurgent groups were able to acquire vast support in the countryside and native areas because of their placement of land reform as their top priority. The white administration, realizing it was vastly outnumbered by the natives, combined with the reluctance to continue fighting a costly war, decided to come to the negotiating table. The United Kingdom also feeling backlash over events in Rhodesia, brokered the negotiations.

Lancaster House Agreement and Resettlement

Debate in Zimbabwe circles after the war revolved around socialism and egalitarianism. Activist and the newly established Zimbabwean African National Union- Patriotic Front, created from its absorption of ZAPU, were committed to building a new socialist society. They saw land as the primary symbol of their struggle (Sachikonye 2003). Nowhere else in Africa had the natives been deprived of so much of this crucial resource (Sachikonye 2003). The notion of land as liberation became cliché in the new black ruling class, and obsession over this symbol would soon come to trump ambitions for true democracy itself.

Bishop Abel Muzorewa would become the first black prime minister of Rhodesia in 1979 (Gregory 1980). Despite the reform, alongside a more representative government, the sanctions that were placed on the country would not be removed immediately. This forced the Muzorewa administration to rely on the still Apartheid South Africa for economic assistance (Gregory 1980). The new president also inherited the mostly white Rhodesian armed forces (Gregory 1980). Both of these factors alienated the president from his initial African nationalists backers.Abel Muzorewa would not last long as president of the country.

Later that year, the British, Rhodesian Government, and African Nationalist came together for negotiations. It would be the first time all three would meet (Gregory 1980). Under the terms of the agreement, land ownership inequities were acknowledged, but transactions could only take place on willing buyer, willing seller terms, which slowed the process (Palmer 1990). Under the agreement, whites who wished to keep their land could do so, and could not be forced out. The British would pay for half of the cost of the resettlement program, and only lands that were considered underutilized were to be acquired by the natives. In a rushed amendment, the parties agreed that the currency used to compensate the farmers had to be remittable in foreign currency (Palmer 1990).

Black peasant farmers had been resilient under the Rhodesian administration despite their condition. They produced maize, which was mainly consumed by the home market, while the white farmers grew more export crops such as tobacco (Palmer 1990). With Rhodesia's declaration of independence in 1965 however, sanctions hit the country, and exports were limited. With this, the white farmers shifted their production to the home market, thus undercutting the small-crop black farmers (Palmer 1990). This made the white farmers and even more important component to the economy of the new Zimbabwe. By 1980, white farmers were producing about "90 percent of the country's commercial food requirements" (Palmer 1990, 167). ZANU-PF had won the elections in 1980 with a resounding victory. The new Zimbabwean government under Mugabe was thus advised not to agitate this key component of the country's economy, and keep resettlement at a gradual pace. That advice appeared to be heeded at first, but was eventually ignored.

By 1990, only about 2 million acres of land were acquired by the Zimbabwean government, only 19 percent of which was considered of prime agricultural quality (PBS 2016). Overall, land transfers were slow, with the Government blaming the restrictions in the Lancaster House Agreement (Sachikonye 2003). Specifically the Zimbabwean government thought that the "willing seller, willing buyer" clause of the negotiation limited potential land supply in comparison to the large demand. Land prices skyrocketed during the post-independence period of the 1980s, and it was doubtful that the Zimbabwean government would have been able to fulfill its duties of resettlement funding (Sachikonye 2003).

Military adventures in the 80's by the Zimbabwe government took energy and effort away for the government's quest of land acquisition. The government first was in the middle of eliminating its political rivals in the north of the country, on the path to achieving its stated objective of a one party state (Krieger 2007). Also, the ZANU-PF government would be involved in Mozambique's civil war (Krieger 2007). Both of these campaigns dried up funds for land reform (Kreiger 2007). The government missed out on prime opportunities to make meaningful land transfers. Under 1985 legislation, farmers in Zimbabwe who wished to sell their land had to give the government the first option (Krieger 2007). Of the 1,800 properties that were available between 1985 and 1992, the government would purchase less than a third of them (Krieger 2007).

There was also little political will for land reform in this period. ZANU-PF would go on to win comfortably in the 1990 and 1995 elections. With the opposition parties at this time were divided, the ruling government had little incentive for substantial action. More attention was paid to the new Economic Structural Adjustment Program (ESAP) prescribed by the International Monetary Fund (IMF). Privatization and trade liberalization would take place in Zimbabwe in the 90s. (Sachikonye 2003). This initially brought promise of increased prosperity for the black population, but instead became the context for endemic corruption at the top of the Zimbabwean Government.

In 1997, citizens, especially war veterans of the Bush War (i.e., against the white Rhodesian government) who were promised generous compensation packages, grew restless. In 1989, the Zimbabwe Liberation War Veterans' Association was established, this brought together former combatants of the conflict to lobby the government for financial assistance (Human Rights Watch 2002). Several laws were passed in the veteran's favor, including the War Victims Compensation Act. However, the administering of these payouts were tainted by corruption, with several high ranking party officials suspected of taking the payments for themselves (Human Rights Watch 2002).

President Robert Mugabe and ZANU-PF tapped into this populist sentiment further by initiating the Land Acquisition Act. The program envisioned the acquisition and transfer of 50,000 of the 112,000 sq. km owned by white farmers (Krieger 2007). The new ruling party, ZANU-PF was determined to go ahead with the land transfers regardless of constitutional constraints. Amendments were added to the constitution to make compensation payments for the white farmers "fair" instead of "adequate" and "within a reasonable time" instead of "promptly" (Krieger 67, 2007). Authority over the matter was also transferred from the courts to the executive, who then had sweeping powers as a result. President Mugabe would frequently state that, "land was a political issue, and would not be derailed by the courts" (Krieger 67 2007). Enabled by their government's tough rhetoric on land, peasants began invading and sacking white owned farms, suspected by many with the passive support of ZANU-PF officials. With public fervor at a high, and the government continuing to blame the law getting in the way in its quest for land transfers, "fast-track" land reform would begin starting in 2000.

Fast Track Reform

The Zimbabwe government in 2000 would put up for referendum a clause into its constitution making the former colonial power (Britain) responsible for all compensation payments, and if those payments were not met, the Zimbabwean government would not be responsible (Krieger 2007). To the ruling party's surprise however, the referendum would fail at the polls. It was a striking defeat for ZANU-PF, a party that prided itself on its populist connections. With the opposition Movement for Democratic Change (MDC) making headway among the populace, ZANU-PF saw the need to preserve its base of support. More sackings of white-owned farms were orchestrated by party officials and the country's Central Intelligence Organization. Squatters and unemployed youths established "bases" on the farms, with the purpose to prevent the MDC from gaining more rural support (Moyo and Yeros 2005).

Ignoring the Supreme Court's objection, the government officially launched its fast track program in July 2000, with the goal of confiscating 9 million more hectares of land. Two types of land transfer models were developed, the "A1" for black peasant farmers, meant to decongest the rural lands, and the "A2" for black commercial farmers (Moyo and Yeros 2005). Derelict, foreign-owned, and underused lands were made priority by the government for transfer, but as more lands were added to the list, the criteria was largely abandoned and the process became arbitrary. Observers have inferred a split within ZANU-PF of those who were in favor of an orderly transition, and officials who preferred more "revolutionary" means (Human Rights Watch 2002). An even more noticeable split could be seen between the ruling party and the national courts.

Violence plagued the Fast Track land reform process. War veterans and ZANU-PF militia often took out their frustration on white farmers, sometimes killing them. In April 2000, a white farmer by the name of David Stevens was shot at point blank range by settlers (Human Rights Watch 2002). Another farmer, Martin Olds, was shot by invading ZANU-PF militia after being lured out of his house which was set ablaze. His mother would also be killed only months later (Human Rights Watch 2002). White farmers have sometimes reacted with violence themselves, exemplified in a 2001 incident where a farmer deliberately ran over a settler with his truck, killing him (Human Rights Watch 2002). Black laborers working for white landowners also felt the wrath of violence. With many white farmers supporting the opposition Movement for Democratic Change party, their black workers were perceived to be complicit by war veterans and ZANU-PF militia. The workers were often intimidated or beaten as a result (Human Rights Watch 2002).

Discrimination in land redistribution on part of the Zimbabwean government was evident. It was widely believed among the populace that support for the ruling ZANU-PF was a criterion for obtaining land (Human Rights Watch 2002). Many residents also said they were ready for new land, but feared applying because of the violent control of the war veterans over the distribution process (Human Rights Watch). Official structures in the country also failed to show adequate control over these veterans. Instances occurred where local governments ordered police to evict squatters and occupiers on farm lands, only for those same police to back down to the demands of the war veterans (Human Rights Watch). Many fast track plots in Zimbabwe ended up going to police, army, and civil servants, taking potential lands away from the poorer peasants (Kreiger 2007).

Because of the haste in which the Zimbabwean government carried out the land reform process, uncertainty about titles and ownership confronted the new owners. Many resettled blacks claimed that they refused to even take up their new property because they did not have the resources to cultivate, and government support was non-existent (Human Rights Watch 2002). Adding to this, communal lands in the hinterland, where many blacks came from, were required to be given up at the time of their new land purchase. But without a secure title elsewhere, some commoners faced the possibility of being landless Human Rights Watch 2002)..

The lack of official ownership also meant that the new owners found it almost impossible to receive the necessary capital from banks to make their agricultural ventures viable. No incoming equipment also meant the new farmers were sitting idle, and doing anything but farming. Stories emerged of new war veteran settlers ignoring their crops, consuming large amounts of alcohol, and continuously slaughtering their livestock for food. A resident said: "On some farms they are planting maize and sunflowers and beans, but on the farm where I work they are just sitting there causing problems. Sometimes they stop the workers going to work; sometimes they are killing cattle to feed themselves" (Human Rights Watch 35). Agricultural production in the country would decline sharply, having severe consequences.

Economic Effects

The effects of this policy on the Zimbabwean economy were overwhelmingly detrimental. Before land reform, farmers had no issue raising funds and obtaining the proper, modern equipment. The primary beneficiaries of land reform were government workers and their families, many of which had no experience with farming. The result was a country that was once self-sustaining in food production, then forced to import large quantities. The resulting collapse in the agricultural sector, which was the country's main economic fixture, crashed the financial and living standards of most Zimbabweans, a disaster that the country is still trying to recover from today.

Because of disruption in production, agricultural output in Zimbabwe fell from 18 percent in 2000 to 14 percent in 2002 (World Bank 2008). The virtual elimination of its agricultural base threw Zimbabwe into an economic tailspin. The government could also no longer meet its debt obligations, resulting in termination in relations with the World Bank, and a hardline isolationist, anti-western stance. Zimbabwe was slowly cut adrift from the Foreign Exchange Market (commonly known as Forex), which enables a country to participate in international trade through currency conversion. The decline in Forex inflow from the decline in exports meant that the country could not import necessary raw materials, which crippled production further.

Zimbabwe also payed a financial price for the political violence that occurred during the land reform process. In reaction to human rights violations, the European Union placed Zimbabwe under sanctions (Kairiza 2012). Under the terms of those sanctions, financial support for all projects in the country effectively ended except for those in "direct support of the population (Kairiza 2012, 8). Additionally, 20 Zimbabwean government officials were banned from travelling to the European Union (Kairiza 2012). Assets that the individuals had in Europe were also seized. Funds from overseas donor agencies also dried up. All of these factors contributed further to the country's economic isolation, and the west was forced to utilize South Africa to influence the Mugabe government.

The economic collapse triggered the situation that Zimbabwe is known for today, hyperinflation. The Zimbabwean government turned to the printing press often after 2000, especially to pay for the cost of land reform (Huchu 2016). Uncontrolled spending and a costly intervention into the Democratic Republic of the Congo, were secondary blows to the nation's economy. The Zimbabwean currency rapidly lost its value, and prices for basic goods would often double every few hours (Huchu 2016). Inflation rates spiraled out of control in the 2000s with the peak rate in 2008, a staggering 79 million percent (Hanke 2009).

The resulting Hyperinflation devastated Zimbabwe's population. Many citizens went hungry during the peak of the crisis due to shortages of food in the markets. Markets at the time were forced to sell at government reduced prices, meaning they could not make profits (Khairiza

2012). Consequently, stores reduced the amount of food on the shelves. (Kairiza 2012). Black market activity increased as it became the only option for people to obtain basic necessities. Many people resorted to bartering, from food, to their farm animals (Kairiza 2012).

The Zimbabwean dollar was so worthless that many domestic businesses pleaded for the government to permit them to use foreign currency. US Dollar's and South African Rand's were already being used at the micro level (Kairiza 2012). In 2009, these private businesses finally got their wish (African Economic Development Institute 2009). This shift to foreign money saw the government license 1,000 shops to sell products in outsider currency (Kairiza 2012). However, Zimbabwe's public sector workers would receive the short end of the deal. Still being paid in the worthless Zimbabwean currency, many public workers could no longer afford public transport, and are close to starvation (African Economic Development Institute 2009). Recently, trade unions have lobbied for government workers to be paid in foreign currency "Numerous teachers, doctors, and policemen have opted to strike, claiming that it was more expensive to work than not to work" (African Economic Development Institute 2009, 3).

With no prospects of an economically viable or socially stable future, many Zimbabwean's left the country. The majority of emigrants left for South Africa, some for Botswana, others for Namibia (Hunchu 2016). This loss of mostly working age populations has caused a "brain drain" in the country. Zimbabwe lost many of its college graduates, young professionals, and former government workers to more prosperous pastures.

Boserupian Theory

In contrast to Malthusian views on population checks and environmental determinism, Danish economist Ester Boseup theorized that humans can overcome their environmental challenges with technology (Marquette 1997). Boserup states that initially, sparsely populated areas will rely on long fallow systems for agricultural production. As the population increases, and resources become more strained, the population will turn to more intensive, technologically based agriculture (Marquette 1997). This thinking is of particular relevance to African continent, which has been historically a sparsely populated area, but is now seeing an explosion of population growth as of recent.

"Boserupian" theories touch on the issue of land and agricultural productivity in Sub-Saharan Africa. She counters the popular notion that much of Africa's land is unproductive and harsh. Zimbabwe's aforementioned productivity disproved this. Rather than the environmental determinist view, she claims that a lack of investment in the proper technology is holding African nations, like Zimbabwe, back (Marquette 1997). In relevance to the Zimbabwe situation, Boserup also states that food imports from the developed world can lead to a death spiral of agricultural productivity and decline (Marquette 1997). Investment in rural, domestic-destined agriculture has decreased. As a result, these areas are less economically viable, and in turn, less attractive to live in compared to the city. The subsequent loss of human capital from these areas makes them less productive, and in turn makes the nation more dependent on imports (Marquette 1997).

Boserupian theory is of particular use to Zimbabwe because of its heavy focus on society and its use of land. It states above all that technology is the key to having a strong and stable agricultural base (Marquette 1997). Boserup also mentions that the largely idle, long fallow systems must be put to use (Marquette 1997). Zimbabwe now is facing a food crisis, with more of its population on the verge of starvation than ever before (African Economic Development Institute 2009). It is therefore important for the country to utilize as much land as sustainably possible to feed its population. With chronic food scarcity, economic growth in the country will be difficult.

Conclusion:

Land is Zimbabwe is plentiful and rich. It has always been the country's most contested resource. The white British settlers of the area exploited it to full use and used it to hold power in the country for almost a century. The excluded majority black population refused to accept this state of affairs and violently overthrew their white overlords. The black administration's policy of land reform was to be the principal aim of change in the country. Done gradually, it could have been achieved with less social and economic cost. But the bitterness of past oppression and the rush to see their new black socialist utopia realized, led ZANU-PF to act in haste. The rush to redistribute land has had lasting social, economic, and health consequences on the country. It is only now that Zimbabwe is starting to emerge from its deep hole. Responsible leadership that utilizes practical measures to benefit every citizen of Zimbabwe, is the only way forward for the country.

To restart its agricultural productivity and boost economic growth, Zimbabwe will likely need help from the outside. This is of course difficult because of the country's political isolation. The country must first make a transition to more democratic politics. Progress on this front has been seen briefly in the form of Mr Morgan Tsvangirai from the opposition MDC being elected Prime Minister in 2008. However a government referendum was held in 2013, with the position of Prime Minister being abolished soon after. Robert Mugabe was re-elected in a general election later that year. Jesse Jackson would comment in 2006 on the issue of land reform, "The process has to attract investors rather than scare them away. What is required in Zimbabwe is democratic rule, democracy is lacking in the country and that is the major cause of this economic meltdown" (Zimbabwe Situation 2006).

Currently, organizations such as the Zimbabwe Agricultural Development Trust (ZADT) have been spearheading efforts to empower smallholder farmers in Zimbabwe. Since loans to these farmers are considered risky by major financial institutions, the smallholders often do not have adequate finances to make a living (ZADT 2016). ZADT has filled in this gap by providing small loans to ":value chains" which the peasant farmers play a significant role (ZADT 2016). "Value Chains" simply refer to the process of the extraction of the food to its placement in a market. It would also be necessary for Zimbabwe to secure larger investments for 21st century, mechanized, farm equipment, which would dramatically improve productivity. Higher Education programs focused on Agricultural and Environmental Science can give the new generation of Zimbabwean farmers the knowledge they need to be productive.

To reboot agriculture in Zimbabwe, there also should be an emphasis on efficiency. Zimbabwean's should tailor their choice of crop to the environmental conditions of their region. Maize is the staple of the country, but it simply cannot be grown everywhere. Tobacco was a substantial export of Zimbabwe before land reform, and can still reap large profits for the country (CIA 2016). The country is also capable of growing sizeable portions of wheat, especially in the well watered center of the country (CIA 2016). There are also some areas of the country especially the south, where it is simply not worth it to plant crops (Mafundikiwa 2014). Rainfall in these areas is too scarce, and the soils too dry. It would be more profitable for these farmers to focus on their cattle and goats, which could be used to purchase food. Zimbabwe is more than capable of becoming a breadbasket of Africa. With the right political leadership and pragmatic policies, the country could find itself in that prime position once again.

References

Floyd, Barry N. 1962. "LApportionment in Southern Rhodesia". Geographical Review, Oct. 1962, Vol. 52:4, P. [566]-582; with Maps, Tables.. n.d.

 Floyd examines land distribution in the colonial era of Zimbabwe (Rhodesia). The author emphasizes how the black population was deprived on land and wealth.

Galbraith, John S. Crown and Charter: The Early Years of the British South Africa Company. Berkeley: University of California Press, 1974.

Galbraith explains the early days of the British South Africa Company. He also examines relationships between members of the board of directors, as well as Cecil Rhodes, the leader of the company. Galbraith claims that Rhodes' leadership style was egocentric and reviled by many of his colleagues.

Huchu, Gladys. Academia.edu - Share Research. Accessed February 21, 2016. http://academia.edu.

Huchu examines the aftermath of Zimbabwe's land reform policy, and its economic effects. The issue of hyperinflation is also examined.

Human Rights Watch. "Fast Track Land Reform In Zimbabwe." 14, no. 1 (March 2002).

Human Rights Watch examined human rights violations in Zimbabwe during the land reform process. Details of notable incidents are included.

Kriger, Norma J. "Liberation from Constitutional Constraints: Land Reform in Zimbabwe." SAIS Review 27, no. 2 (2007): 63-76. doi:10.1353/sais.2007.0034.

Kreiger examines the legality of the Zimbabwean governments land reform strategy. The author claims that the government often ignored its own laws it its forced confiscation of white owned farms.

Moyo, S., and P. Yeros. 2005. "Land occupations and land reform in Zimbabwe: towards the national democratic Revolution". Reclaiming the Land : the Resurgence of Rural Movements in Africa, Asia and Latin America.. 65.

Moyo and Yeros report on the social inequities of land distribution in Zimbabwe, and how the issue is common in other countries. The authors claim that the government's intentions were correct and land reform was an absolute necessity for the welfare of the country.

Palmer, Robin. "Land Reform in Zimbabwe, 1980-1990." African Affairs 89 (April 1990).

Palmer examines the chronology of land reform in Zimbabwe from 1980 to 1990. Details of legislation created by the Zimbabwean government as basis for its land redistribution are included.

"PBS: Public Broadcasting System | 404." PBS: Public Broadcasting Service. Accessed March 10, 2016. http://www.pbs.org/newshour/bb/africa/land/gp_zimbabwe.html.

PBS reports on Zimbabwe's land inequality, and the effects of the subsequent land reform. Social and economic effects of the policy are examined.

Sachikonye, Lloyd M. "From 'Growth with Equity' to 'Fast-Track' Reform: Zimbabwe's Land Question." Review of African Political Economy 30, no. 96 (2003): 227-240. doi:10.1080/03056244.2003.9693496.

Sachikonye emphasizes Zimbabwe's economic history after 1990. Economic events in the country are examined chronologically. The author emphasizes that lack of fiscal discipline, land reform, and military interventions triggered the country's current crisis.

Zvobgo, C. J. M. A History of Zimbabwe, 1890-2000 and Postscript, Zimbabwe, 2001-2008. Newcastle upon Tyne, UK: Cambridge Scholars Pub, 2009.

Zvobgo gives a general history of Zimbabwe from 1890-200. The report details events in the nation from the start as a colony, to an independent, minority rule state, and finally in its current form as Zimbabwe.

CPSIA information can be obtained
at www.ICGtesting.com
Printed in the USA
LVIC06n0031170418
573763LV00005B/24